WEB DEVELOPER

A Crabtree Branches Book

B. Keith Davidson

CRABTREE
Publishing Company
www.crabtreebooks.com

School-to-Home Support for Caregivers and Teachers

This high-interest book is designed to motivate striving students with engaging topics while building fluency, vocabulary, and an interest in reading. Here are a few questions and activities to help the reader build upon his or her comprehension skills.

Before Reading:

- *What do I think this book is about?*
- *What do I know about this topic?*
- *What do I want to learn about this topic?*
- *Why am I reading this book?*

During Reading:

- *I wonder why...*
- *I'm curious to know...*
- *How is this like something I already know?*
- *What have I learned so far?*

After Reading:

- *What was the author trying to teach me?*
- *What are some details?*
- *How did the photographs and captions help me understand more?*
- *Read the book again and look for the vocabulary words.*
- *What questions do I still have?*

Extension Activities:

- *What was your favorite part of the book? Write a paragraph on it.*
- *Draw a picture of your favorite thing you learned from the book.*

TABLE OF
CONTENTS

IN MY COMMUNITY

A community is made up of people coming together to make life better for each other.

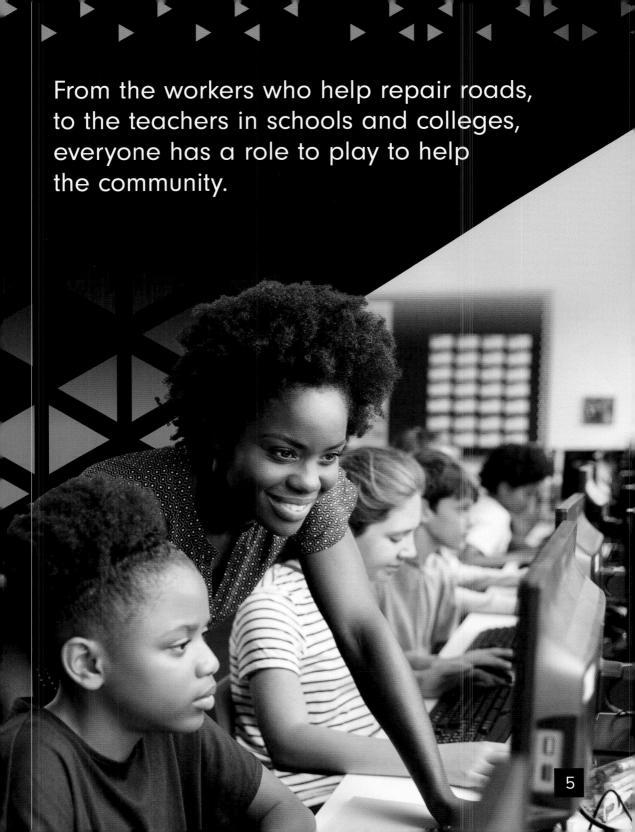

From the workers who help repair roads, to the teachers in schools and colleges, everyone has a role to play to help the community.

Web developers are important members of our **global** community.

6

They design the web pages that we look at every day. They write the code that tells the website how to function.

```
redrawSiteWithMagic(){
(window).width() < 700){
'.rotate-left').addClass('rotate-left-di
'.rotate-right').addClass('rotate-right-d
rotate-left').removeClass('rotate-left')
rotate-right').removeClass('rotate-right

rotate-left-disactivated').addClass('rot
rotate-right-disactivated').addClass('
rotate-left-disactivated').removeClass('
rotate-right-disactivated').removeClass(
```

WEB DEVELOPER

Many web developers have degrees or diplomas in a computer-related field. These certifications are not requirements, but they will help start a career.

Many developers go through bootcamps— weekend courses designed to teach you everything you need to know in a very short period of time.

What's the difference between a web developer and a computer programmer? Computer programmers design software and write code. Web developers write code, but for websites only.

SPECIAL SKILLS

A basic knowledge of computers and how they operate will help, but websites have a language all their own.

You will have to learn the basic computer languages of HTML (hypertext markup language), CSS (cascading style sheets), and **JavaScript**. These are the three most common languages in website development.

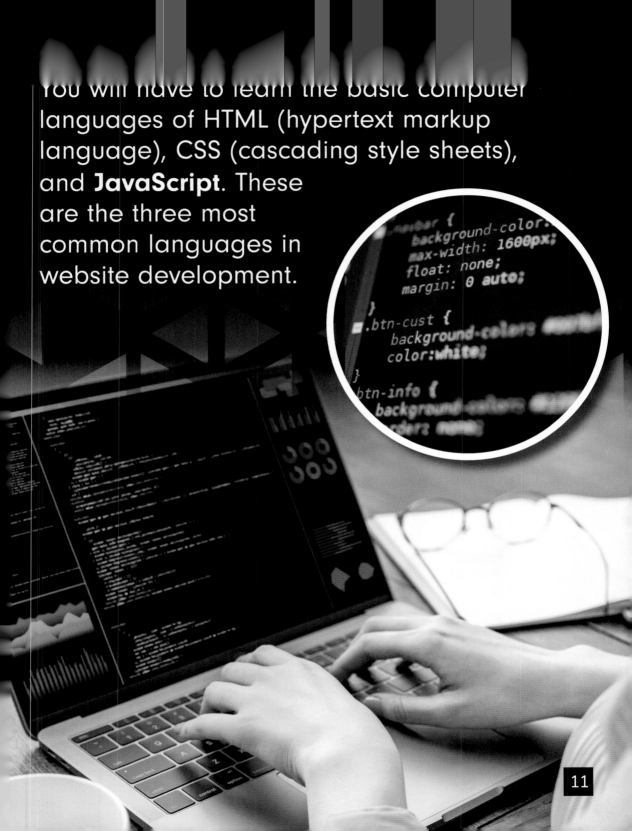

Some developers choose to work on **front-end** web development—the part that you view on your computer screen.

Back-end developers work on the databases and **server** side of the internet. They focus on the collection of information. **Full-stack** developers work on both sides of the website.

Web developers
work with their
clients to design
the perfect website
for a business,
a retail chain, a
government, or
a charity.

Each client will have different needs. The developer must be able to create a website that works for each of them.

Two thirds of web usage is now done on smartphone screens. So developers have to ensure that their websites work well on mobile devices.

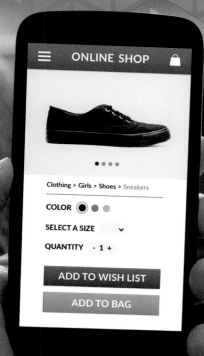

WORKING CONDITIONS

Web developers work in offices, **cubicles**, or in their own homes. If the computer they're using is a laptop, they can work almost anywhere.

All they need for their work is a computer, Internet access, a web **browser**, and web development tools.

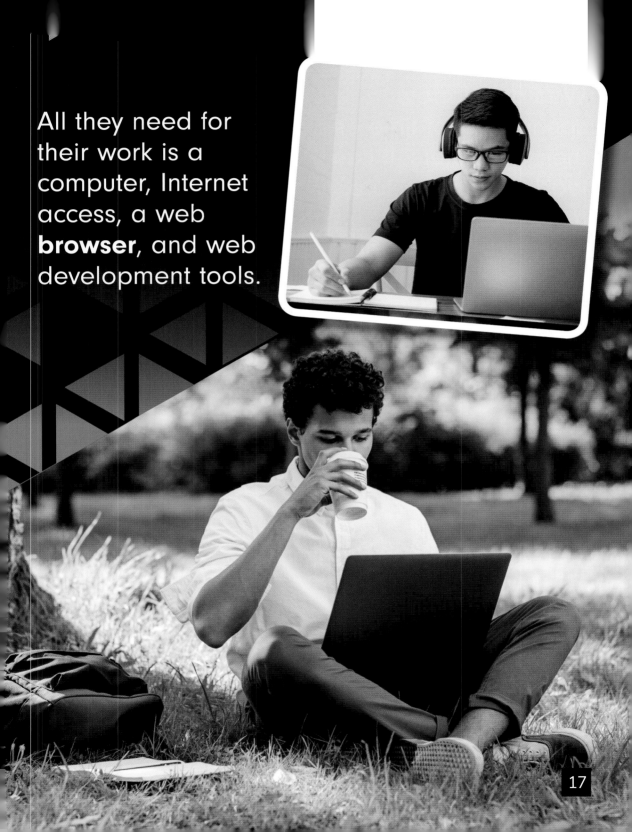

Web developers spend a lot of time on their computers. It can lead to poor posture, muscle aches, and injuries. Work stations must be set up properly so work can be done comfortably.

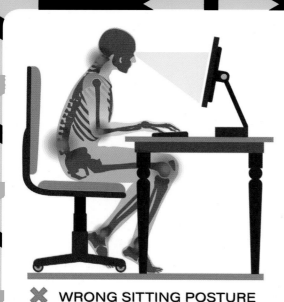

❌ WRONG SITTING POSTURE

✔ CORRECT SITTING POSTURE

Many web development companies have gyms and other activities available for their employees to keep their bodies moving during long, busy days.

Knowing the coding languages is only half the job. Web developers work closely with their clients to make sure their design meets the customer's expectations. Developers often have to be coders, designers, AND salespeople.

The lifespan of a website is about three years. Technology is advancing so quickly that some websites may stop working properly within that time.

Web developers have to think about their clients, their clients' customers, and their competitors. There are billions of websites out there on the web. Developers have to find ways to grab attention and get views.

The source code of a website can affect how it is ranked by search engines. Google ranks websites based on certain HTML elements.

THE CHALLENGES YOU WILL FACE

Not all computer languages are created equal. They all have their issues. Developers have to know which languages are better suited to deliver the website their client needs.

Another challenge is the ever-changing technology. Every year there are new devices, applications, and programs. A web developer has to stay on top of all of it.

A REWARDING CAREER

A web developer connects people and businesses to the world.

It is rewarding work that comes with a good salary, though the salaries vary by industry.

Publishing	$101,905 - $123,870
Computer Systems	$68,450 - $75,500
Advertising	$65,370 - $71,700
Scientific and Technical	$63,950 - $70,790

Being a web developer allows you to shape the way people see the world around them. It is a fast-paced and exciting career that is constantly growing and changing.

It is a good career path for anyone who loves working with computers and the Internet.

GLOSSARY

back-end (BAK-end): the part of the website that we don't see that focuses on the server and data collection

browser (BROW-sur): software that lets a user access an information system called the World Wide Web (WWW)

clients (KLY-uhnts): people who pay for the services of others

code (KOHD): program instructions for computers

cubicles (KYOO-buh-kuhlz): a small partitioned area of a room that functions as an office

front-end (FRUHNT-end): the part of the website that people view

full-stack (FUL-stak): refers to the whole website

global (GLOW-bul): relating to the whole world

JavaScript (JAH-vuh-skript): a computer code commonly used for websites

server (SUR-vur): a large computer with more storage and processing power

software (SAWFT-wair): programs designed for computers

source code (SORSS-kohd): the main code used to write a website's code

WEBSITES TO VISIT

scratch.mit.edu

www.bls.gov/ooh/computer-and-information-technology/web-developers.htm#tab-1

https://careerfoundry.com/en/blog/web-development/what-does-it-take-to-become-a-web-developer-everything-you-need-to-know-before-getting-started

ABOUT THE AUTHOR

B. Keith Davidson

B. Keith Davidson has had careers in agriculture, industrial manufacturing, and the service industry. His career in education led to his current career in writing books.

Written by: B. Keith Davidson
Designed by: Jennifer Dydyk
Edited by: Kelli Hicks
Proofreader: Ellen Rodger
Print and production coordinator:
Katherine Berti

Photographs: Cover career logo icon © Trueffelpix, diamond pattern used on cover and throughout book © Aleksandr Andrushkiv, cover photo © PR Image Factory/shutterstock.com, photo at top of cover and top of title page © ronstik/shutterstock.com. Page 4 top photo © David Econopouly | Dreamstime.com, bottom photo © Firmanemmanuelle | Dreamstime.com, Page 5 © Monkey Business Images | Dreamstime.com, Page 7 © Andrey Sirant | Dreamstime.com, All other images from istock by Getty Images: Page 6 © SeventyFour, Page 8 © Photo Italia LLC, Page 9 top photo © monkeybusinessimages, bottom photo © nd3000, Page 10 © milindri, Page 11 inset photo © Minerva Studio, bottom photo © metamorworks, Page 12 © cyther5, Page 13 both photos © SeventyFour, Page 14 top photo © NicoElNino, bottom photo © kasto80, Page 15 © CarmenMurillo, Page 16 © SolisImages, Page 17 top photo © ijeab, bottom photo © Youngoldman, Page 18 top photo © thodonal, illustration © bestsale, Page 19 © howtogoto, Page 20 © NiKita Filippov, Page 21 top photo © Rostislav_Sedlacek, bottom photo © pondsaksit, Page 22 © Wasan Tita, Page 23 background photo © NanoStockk, top photo © Baks, bottom photo © mbbirdy, Page 24 © YurolaitsAlbert, Page 25 © Igor-Kardasov, Page 26 © Berezko, Page 27 © gorodenkoff, Page 28 © scyther5, Page 29 © Deagreez

Library and Archives Canada Cataloguing in Publication

Available at the Library and Archives Canada

Library of Congress Cataloging-in-Publication Data

Available at the Library of Congress

Crabtree Publishing Company

www.crabtreebooks.com   1-800-387-7650

Published in the United States
Crabtree Publishing
347 Fifth Avenue
Suite 1402-145
New York, NY, 10016

Published in Canada
Crabtree Publishing
616 Welland Ave.
St. Catharines, ON
L2M 5V6

Printed in the U.S.A./CG20210915/012022